JAPANESE WEAPONS ILLUSTRATED

1944

The Naval & Military Press Ltd

Published by

The Naval & Military Press Ltd

Unit 5 Riverside, Brambleside
Bellbrook Industrial Estate
Uckfield, East Sussex
TN22 1QQ England

Tel: +44 (0)1825 749494

www.naval-military-press.com
www.nmarchive.com

JAPANESE WEAPONS ILLUSTRATED

CONTENTS

CONTENTS (*Contd.*)

CONTENTS *(Contd.)*

DISTRIBUTION

All formation and unit intelligence officers   ...   ...   2 copies each

OCTUs and Pre-OCTUs ...   .   Scale III

INTRODUCTION

This book is not intended to be a technical handbook, but an aid to the identification of Japanese Army equipment.

Sections 3, Self-propelled guns, and 9, Military bridges, represent classes of equipment which the Japanese Army must be expected to develop, and for which illustrations will be provided as and when they become available.

Early examination by experts of equipment of new or modified designs is of the utmost importance, and anything found but not shown in this book should be reported without delay.

JAPANESE WEAPONS ILLUSTRATED

YEAR DATES OF JAPANESE WEAPONS

1. The Japanese use two systems to denote the year or " model " number of their weapons and equipment.

(a) *System based on year of reigning Emperor*

Under this system the year numbers correspond with the years of reign of the last three Emperors, namely :—

(i) Emperor MEIJI, whose reign began in 1868;

(ii) Emperor TAISHO, whose reign began in 1912;

(iii) Emperor SHOWA, who has reigned since 1926.

To determine the corresponding European date :—

In the MEIJI period, add the Japanese year date to 1867, thus " Model 38 " rifle is that adopted in the 38th year of MEIJI (A.D. 1905).

In the TAISHO period, add the Japanese year date to 1911, thus " Model 11 " LMG is that adopted in the 11th year of TAISHO (A.D. 1922).

In the SHOWA period the second system, as shown in paragraph (b) below, is usually adopted for *model* of weapons. However, the *date* of manufacture of weapons and equipment is frequently given under the SHOWA system. The corresponding European date is found by adding 1925, thus a weapon *dated* " SHOWA 16 " was manufactured in 1941.

(b) *JIMMU system, based on year of founding of Japanese Empire*

Under this system year numbers run consecutively from the legendary founding of Japan under JIMMU, its first Emperor, in 660 B.C. Weapons designed in the SHOWA period (since 1926) are usually numbered on this system.

To determine the corresponding European date, subtract 660 from the Japanese date. Thus the Japanese year 2600 is A.D. 1940, the year 2595 is A.D. 1935, and so on. In practice the number is commonly abbreviated to the last two figures, so that the MMG " Model 92 " is that adopted in 1932, and the medium tank " Model 97 " that adopted in 1937.

The Model 0 aircraft (the Zero) was introduced in 2600 (A.D. 1940), and the 47-mm A tk gun Model 1 was introduced in 1941.

SECTION 1

TANKS

Japanese tank development began with the purchase of tanks from European armament manufacturers, and for some years the Japanese did little more than copy these tanks, introducing modifications from time to time. During the middle 1930's, however, they started designing their own tanks, which show very few signs of outside influence.

The latest tanks in service are a tankette M 2597, a light tank M 2595, a medium tank M 2597, and an amphibious tank. After these, the next in importance is probably an early design of tankette of which there are two models M 2592 and M 2594.

A few heavy tanks have been reported in the past, but the majority of these seem to have been copies of old Vickers designs of the pre-1930 period. Details of a somewhat later 27-ton tank, M 2595, have also been received, but this tank also is believed to be completely obsolete.

PLATE 1.—Carden-Loyd Two-man Tank

An old equipment, dating from about 1928 and believed to have been used by Japanese naval troops in Shanghai. Approximate details: weight: 2 tons. Length: 8 ft 0 in. Width: 5 ft 6 in. Height: 4 ft 6 in. Armament: Vickers type water-cooled MG in fixed superstructure. Crew: **2**.

PLATE 2.—Tankette M 2592 (1932)
Weight : 3 tons. Length : 11 ft. Width : 5 ft 3 in. Height : 5 ft 4 in.
Armament : one MG in turret. Crew : 2.

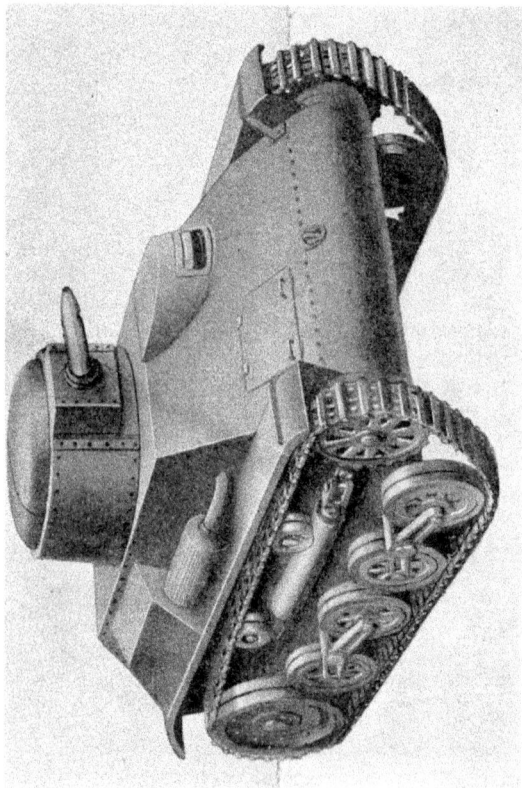

PLATE 3.—Tankette M 2597 (1937)

Weight: **4 tons 3 cwt.** Length: **12 ft.** Width: **6 ft.** Height: **6 ft.**
Armament: either one 37 mm (1·45 in) gun or one MG. Crew: **2.**
Illustration shows turret MG in protecting sheath. For appearance of
37 mm (1·45 in) gun, *see* light tank M 2595 (Plates 5 and 6).

PLATE 4.—Light Tank M 2593 (1933)

An early light tank, now obsolescent. Approximate details: Weight: 7 tons. Length: 14 ft 8 in. Width: 5 ft 11 in. Height: 6 ft 0 in. Armament: two MGs   Crew: three.

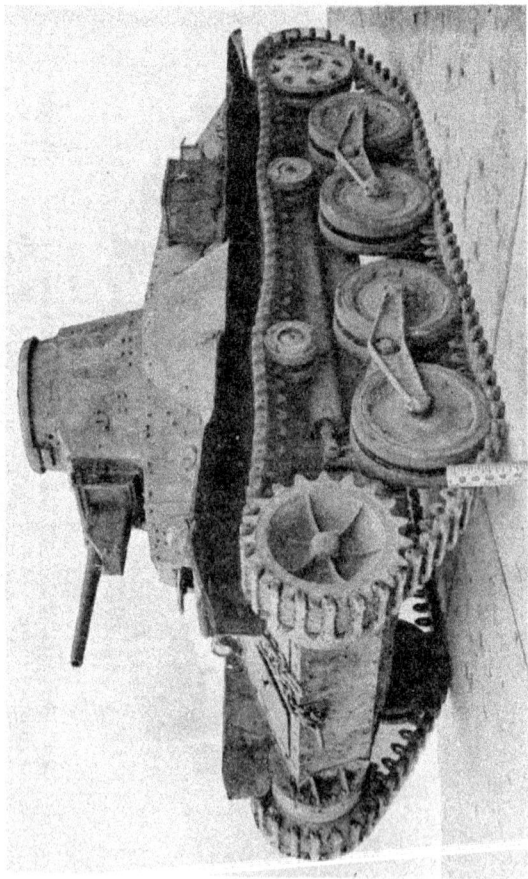

PLATE 5.—Light Tank M 2595 (1935)—nearside view

Weight: 7–9 tons. Length: 14 ft 4 in. Width: 6 ft 9 in. Height: 7 ft 0 in. Armament: one 37 mm (1·45 in) gun and two 7·7 mm (·303 in) MGs. Crew: 3.

9

PLATE 6.—Light Tank M 2595 (1935)—offside view
One MG is mounted in the offside rear of the turret and the other in the nearside front of the superstructure.

PLATE 7.—Amphibious Tank

Weight: 9 tons 8 cwt. Length: 15 ft 10 in. Width: 9 ft 2 in. Height: 7 ft 8 in. Armament: 37 mm (1·45 in) gun and two MGs. Crew: 5. The detachable pontoons are not included in the above particulars.

PLATE 8.—Medium Tank M 2594 (1934)—offside front view

The latest of three models developed from the Vickers Mark C, but now being superseded by the M 2597 medium tank. Weight: 14 tons. Armament: 57 mm (2·24 in) gun and two MGs. Crew 5

PLATE 9.—Medium Tank M 2594 (1934)—nearside rear view

Note ditch-crossing tail, and mounting for MG in rear of turret. Approximate dimensions are : Length : 20 ft 9 in (including tail). Width : 8 ft 4 in. Height : 8 ft 6 in.

PLATE 10.—Medium Tank M 2597 (1937)

Weight in action : 15 tons (approx). Length : 18 ft. Width : 7 ft 8 in.
Height : 7 ft 8 in. Armament : one 57 mm (2·24 in) gun and two MGs.
Crew : **4**. Note railing type aerial on turret.

PLATE 11.—Medium Tank M 2597 (1937)
Same tank as in Plate 10, but with railing type aerial removed. This
had been removed before capture.

SECTION 2

ARMOURED CARS

No Japanese armoured car of a type developed since 1933 has yet been identified.

PLATE 1.—4-wheeled armoured car—Vickers-Crossley type

Weight : 4¾ tons. Length : 16 ft 5½ in. Width : 6 ft 1½ in. Height : 8 ft 5½ in. Armament : two MGs which can be used in any two of four mountings in turret. Crew : 4.

PLATE 2.—4-wheeled armoured car—Osaka M2592 (1932)
Weight : 5¾ tons   Length : 16 ft 5 in.   Width : 6 ft 0 in.   Height : 8 ft 8 in.   Armament : two MGs (one in turret).   Crew : 4.

27

PLATE 3.—6-wheeled road-rail armoured car—Sumida M2593 (1933)
Weight: 7 tons. Length: 21 ft 6 in. Width: 6 ft 3 in. Height: 9 ft 8 in. Armament: one MG in turret, also seven pistol or machine carbine ports in hull. Crew: 6.

SECTION 3

SELF-PROPELLED GUNS

SECTION 4

ANTI-TANK GUNS

The development of Japanese anti-tank guns has, by European standards, been slow. It is possible, however, that heavier equipments exist, and German influence on new design may become apparent.

PLATE 1.—37-mm (1·45-in) A tk gun. Model 94 (1934)

Muzzle velocity : 2,300 fs. Weight of AP shell : 1 lb 8½ oz. Estimated perforation :   36-mm (1·4-in) at 250 yds, 32-mm (1·26-in) at 500 yds at 30 degrees impact.

PLATE 2.—37-mm (1·45-in) A tk gun.   Model 97 (Rheinmetall)
Muzzle velocity : 2,625 fs.   Weight of AP shell : 1 lb 10¾ oz.   Estimated perforation : 41-mm (1·6-in) at 250 yds, 36-mm (1·4-in) at 500 yds at 30 degrees impact.

PLATE 3.—47-mm (1·85-in) A tk gun. Model 1 (1941)
Muzzle velocity : 2,701 fs. Weight of AP shell : 3 lb 6 oz. Estimated perforation : 59-mm (2·4-in) at 500 yds, 45-mm (1·8-in) at 1,000 yds at 30 degrees impact.

SECTION 5

ARTILLERY

The Japanese field artillery presents a somewhat mixed picture since a great variety of models is in use, considering the small number of different calibres involved. The most modern types in use are the 75 mm mountain gun Model 94 (1934), the 105 mm howitzer Model 91 (1931), the 105 mm gun Model 92 (1932), and the 150 mm howitzer Model 96 (1936), but all those illustrated are liable to be met, owing to the apparently slow rate of replacement.

PLATE 1.—70-mm (2·76-in) battalion gun. Model 92 (1932)
MV : 650 fs. Max range : 3,000 yds. Weight of shell : 8·4 lb.

PLATE 2.—75-mm (2·95-in) infantry regimental gun. Meiji 41 (1908)
MV : 1,250 fs. Max range : 7,700 yds. Weight of shell : 14 lb.

PLATE 3.—75-mm (2·95-in) field gun. Model 90 (1930)
MV : 2,230 fs.   Max range : 13,300 yds.   Weight of shell : 14·3 lb.

74

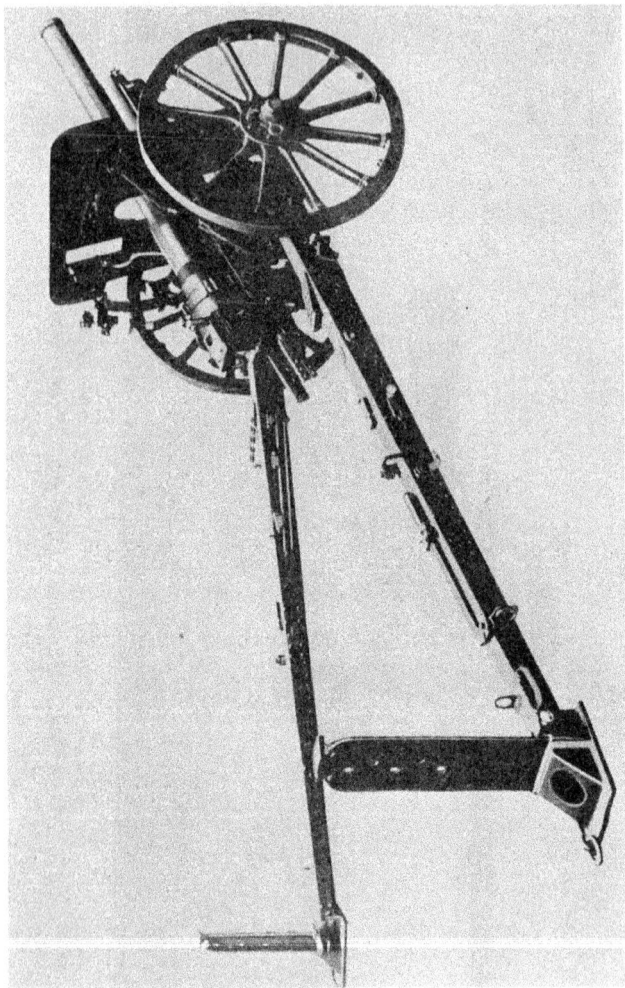

PLATE 4.—75-mm (2·95-in) mountain gun. Model 94 (1934)
MV : 1,300 fs. Max range : 8,750 yds. Weight of shell : 13·4 lb.

PLATE 5.—105-mm (4·14-m) howitzer. Model 91 (1931)

MV : 1,790 fs.   Max range : 11,811 yds.   Weight of shell : 33 lb.

PLATE 6.—105-mm (4·14-in) medium gun. Model 92 (1932)
Max range : 16,400 yds. Weight of shell : 33 lb.

PLATE 7.—150-mm (5·9-in) medium howitzer. TAISHO " 4 " (1915)
MV : 1,350 fs. Max range 9,460 yds. Weight of shell : 80 lb.

PLATE 8.—150-mm (5·9-in) medium howitzer. Model 96 (1936)
Max range : 11,400 yds. Weight of shell : 80 lb.

SECTION 6

ANTI-AIRCRAFT GUNS

Japanese AA equipments of 25 mm (·98 in) and upwards are of original naval design, modified later for land service.

Of the heavy guns, the 75 mm (2·95 in) types are mounted on a single axle carriage and the 105 mm (4·14 in) type on a double axle carriage. A 127 mm (5 in) naval gun (not shown) is a static equipment.

All the heavy guns are dual purpose, AA—Coast defence weapons. The light guns 20 mm (·79 in) and 25 mm, and the 75 mm guns may also be used in the anti-tank role.

PLATE 1.—13 mm (·51 in) AA/A tk MG, Model 93 (1933) (twin mounting)
Weight of complete equipment   7¼ cwt.   Weight of each gun : 87 lb.
Overall length : 89 ins.   Feed   30 round box magazine.   Max vertical
range : 10,800 ft. Ground sights graduated to 3,600 metres. Ammunition :
Semi-rimless ball, tracer, AP.

PLATE 2.—20 mm (·79 in) AA/A tk gun.   Model 98 (1938)

Muzzle velocity : 2,720 fs.   Max horizontal range : 5,450 yds.   Max vertical range : 12,000 ft.   Ammunition fired : HE/T, AP/T.   Practical rate of fire : 120 rpm.

PLATE 3. 20 mm (·79 in) AA gun Oerlikon (Model J Las).

Muzzle velocity : 2,720 fs. Max horizontal range : 5,450 yds. Max vertical range : 12,200 ft. Weight of shell : 4½ oz. Practical rate of fire : 120 rpm.

PLATE 4.—25 mm (·98 in) AA/A tk gun. Model 96 (1936) (twin mounting).
Muzzle velocity: 2,978 fs. Max horizontal range: 5,470 yds. Max vertical range: 12,000 ft. Weight of shell: 8½ oz. Practical rate of fire: 240 rpm.

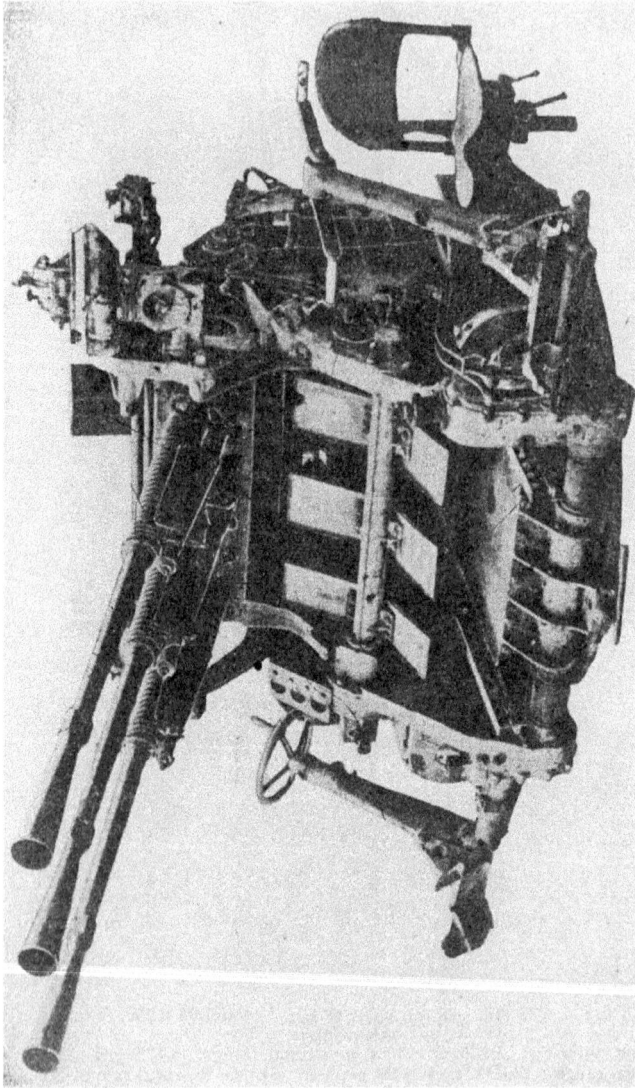

PLATE 5.—25 mm (·98 in) A.A/A tk gun. Model 96 (1936) (triple mounting). Weight of shell : 8½ oz. Practical rate of fire : 360 rpm. Muzzle velocity : 2,978 fs. Max horizontal range : 5,470 yds. Max vertical range : 12,000 ft.

PLATE 6.—75 mm (2·95 in) A.A gun. Taisho "11" (1922).
Muzzle velocity : 1,800 f.s. Max horizontal range : 11,000 yds. Max vertical range : 19,600 ft. Weight of shell : 14 lb 8 oz. Practical rate of fire : 12 rpm. This is obsolescent.

PLATE 7.---75 mm (2·95 in) AA gun.   Model 88 (1928).

Muzzle velocity : 2,400 fs.   Max horizontal range : 15,100 yds.   Max vertical range : 29,000 ft.   Weight of shell : 14 lb 5 oz.   Practical rate of fire : 20 rpm.

PLATE 8.—105 mm (4·14 in) A.A. gun. Taisho " 14 " (1925).

Muzzle velocity : 2,300 fs. Max horizontal range : 19,600 yds. Max vertical range : 36,000 ft. Weight of shell : 35 lb. Practical rate of fire : 8 rpm.

SECTION 7

MORTARS AND GRENADE DISCHARGERS

More originality has been displayed by the Japanese in mortar design than in any other type of weapon. In addition to their standard 50 mm (1·97 in) grenade discharger Model 89 (itself an effective and ingenious weapon), and the standard 81 mm (3·18 in) mortar Model 97, several novel types of mortar have been identified.

The 50 mm Model 98 mortar fires a box-shaped HE stick bomb and a demolition tube against wire and pillboxes at comparatively short ranges and should prove effective in this role. The 70 mm (2·75 in) barrage mortar, which projects a canister containing seven small parachute bombs, is believed to be a weapon of defence against low flying aircraft, but may well be used for other tasks.

The " Small " 81 mm mortar Model 99 is less than half the weight of normal 3 in mortars, but fires a 7 lb bomb to approximately 2,000 yds. A similar weapon has recently been introduced into the German Army.

The 90 mm (3·54 in) mortar is chiefly remarkable for its substantial recoil mechanism and fires both HE and incendiary ammunition.

150 mm (5·9 in) mortars have also been identified but no photographs are available to date.

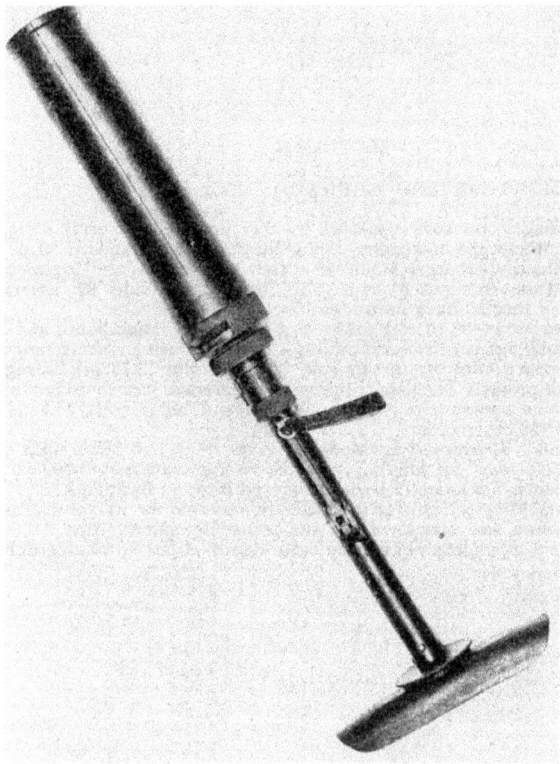

PLATE 1.—50 mm (1·97 in) grenade discharger TAISHO " 10 " (1921)
Weight: 5¼ lb. Overall length: 20 in. Ammunition: Model 91 grenade, weight 18¾ oz. Max range: 175 yds. Smoke and signal grenades are also fired.

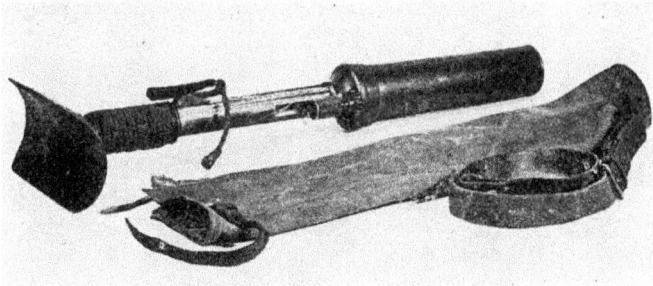

PLATE 2.—50 mm (1·97 in) grenade discharger Model 89 (1929). Weight: 10¼ lb. Overall length: 24 in. Ammunition: (a) Model 89 bomb. Weight: 1¾ lb. Weight HE filling 5 oz. Max range 710 yds. (b) Model 91 grenade. Weight: 18¾ oz. Max range 175 yds. A rifled muzzle-loading weapon.

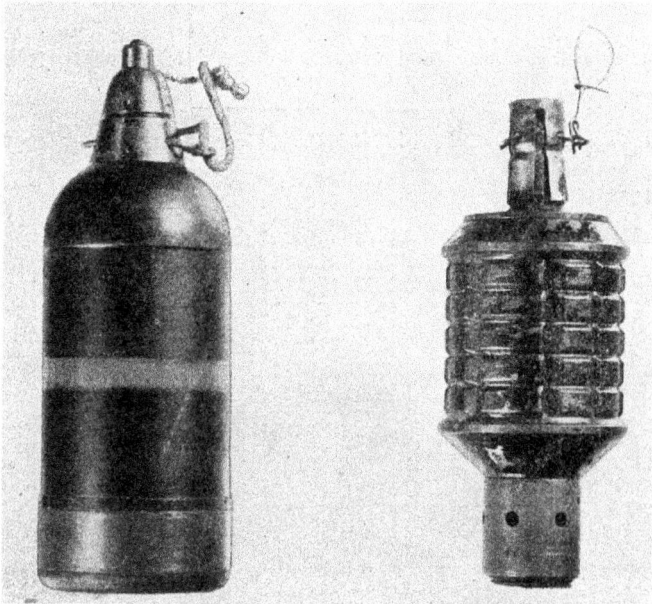

PLATE 3.—Ammunition for 50 mm (1·97 in) grenade discharger. Model 89 (1929). (See Plate 2). Left: Model 89 bomb. Right: Model 91 grenade

PLATE 4.—50 mm (1·97 in) mortar Model 98 (1938).
Total weight : 48 lb. Ammunition : (*a*) HE Stick bomb : Total
weight 14 lb. Weight filling : 7 lb. Max range : Approx 450 yds.
(*b*) Demolition tube. Total weight : 18½ lb. Weight filling : 5 lb.
Max range : Approx 320 yds.

PLATE 5.—70 mm (2·75 in) barrage mortar.
A smooth bore percussion fired weapon firing a projectile containing
seven small high explosive parachute bombs. Length overall : 75 in.

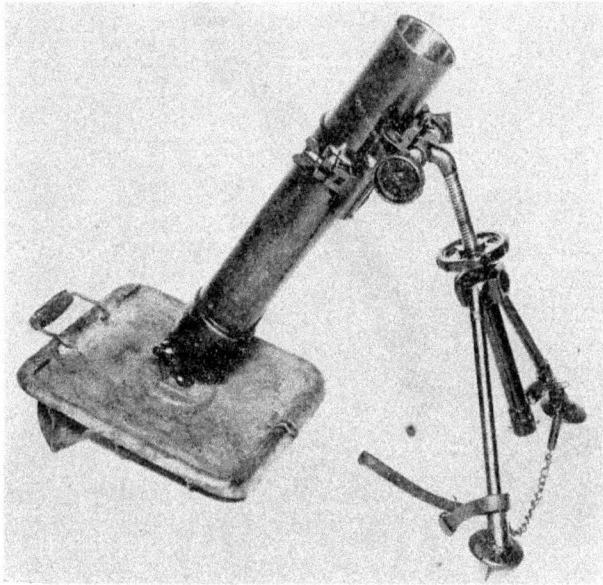

PLATE 6.—81 mm (3·18 in) small mortar Model 99 (1939).

Weight : 52 lb.   Overall length barrel : 25¼ in.   Weight of bomb : 7 lb.
Max range : 2,000 yds (approx).

114

PLATE 7.—81 mm (3·18 in) mortar Model 97 (1937).
Weight : 145 lb.   Overall length barrel : 49½ in.   Weight of bomb   7 lb.
Max range : Approx 3,100 yds.

115

PLATE 8.—90 mm (3·54 in) mortar Model 94 (1934).
Weight : 340 lb. Overall length of barrel : 50 in. Weight of bomb :
11½ lb. Max range : 4,150 yds.

SECTION 8

SMALL ARMS

The Japanese have borrowed extensively from European sources in the design of their small arms, particularly German, French and Czech. In general, however, the range of small arms encountered so far, although adequate by pre-war standards, does not impress when compared with the latest Allied or German weapons. It is probable that more modern weapons are still to be identified and these may show considerable improvements in design, particularly if German help has been enlisted. Only one instance of direct German influence in this field has been met to date, the hollow charge rifle grenade (Plate 5), but more instances may be expected, particularly in anti-tank weapons in which the Japanese are comparatively weak.

One feature of interest is the change over from 6·5 mm (·256 in) to 7·7 mm (·303 in) as the standard small arms calibre. This appears to be taking some time as 6·5 mm weapons are still widely encountered.

PLATE 1.—8 mm (·315 in) Self-loading Pistol, " NAMBU " Model.

Weight : 31 oz. Feed : 8 round magazine in butt. Ammunition : Semi-rimless ball.

PLATE 2.—8 mm (·315 in) Self-loading Pistol, TAISHO " 14 " (1925).
Weight : 2 lb. Feed : 8 round magazine in butt. Ammunition : Semi-rimless ball.

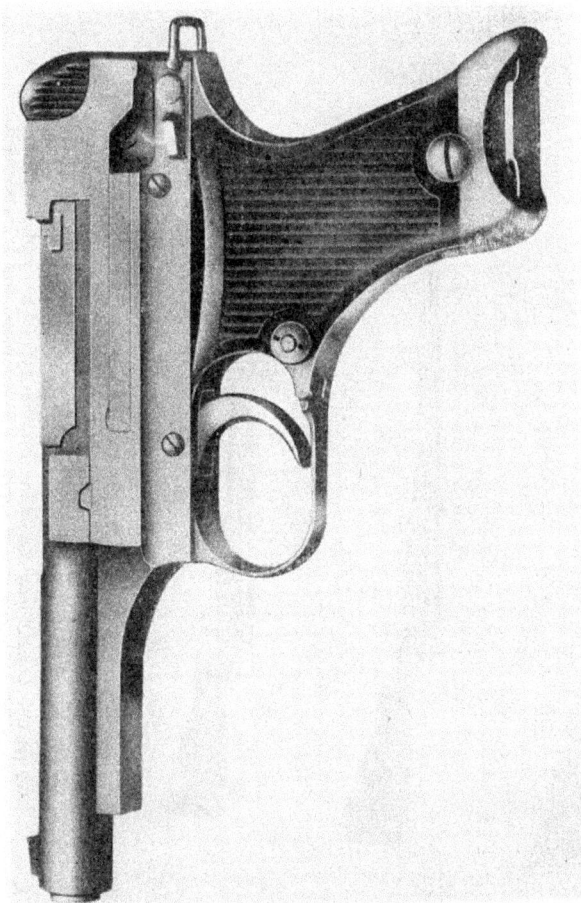

PLATE 3.—8 mm (·315 in) Self-loading Pistol, Model 94 (1934). Weight : 27 oz. Feed : 6 round magazine in butt. Ammunition : Semi-rimless ball.

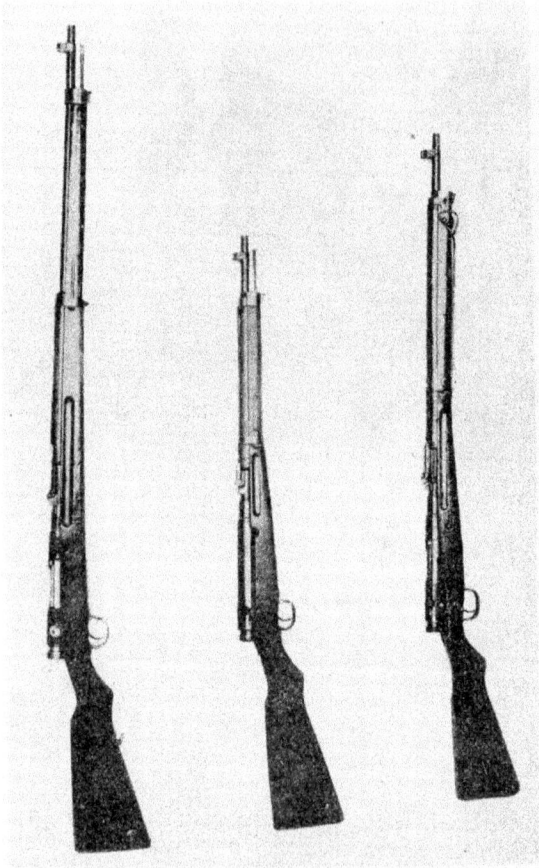

PLATE 4.—Rifles.

Top : 6·5 mm (·256 in) long rifle, Meiji " 38 " (1905). Overall length : 50·2 in. Centre : 6·5 mm (·256 in) short rifle, Meiji " 38 " (1905). Overall length : 38¼ in. Bottom : 7·7 mm (·303 in) short rifle, Model 99 (1939). Overall length : 45 in. Another type of 6·5 mm carbine and a long 7·7 mm rifle are in service. Telescopic sights are also used.

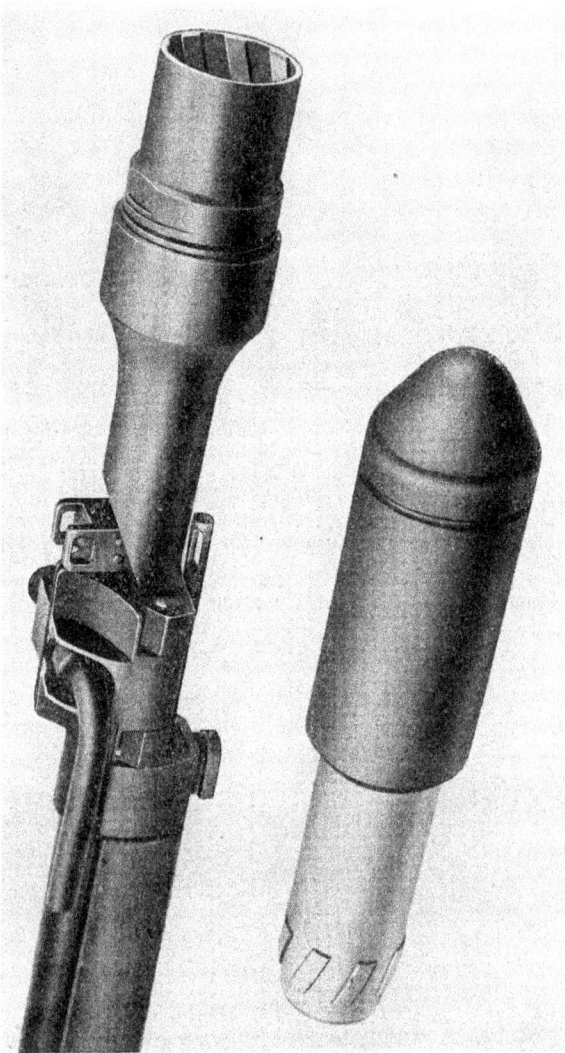

PLATE 5.—Rifle Grenade Discharger and Grenade.
The discharger is fitted to the standard 6·5 mm (·256 in) rifle and fires a hollow charge A tk projectile. A copy of a German weapon.

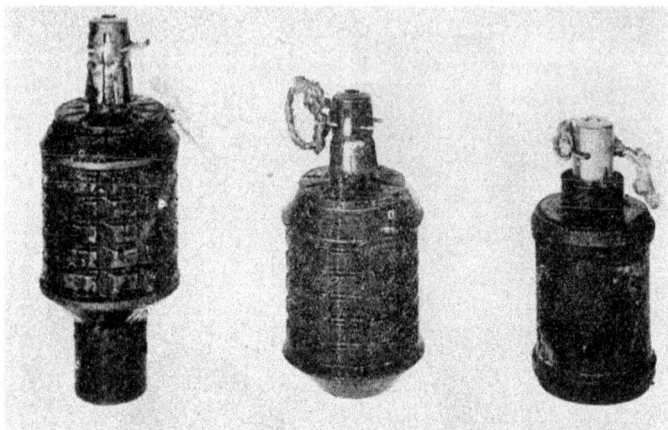

PLATE 6.—Grenades.

Left : Model 91 (1931) grenade. Weight : 18¾ oz. Thrown, fired from
Model 89 grenade discharger or from a rifle spigot discharger in which
case it is fitted with a finned tail unit. Delay : 8–9 seconds. Centre :
Model 97 (1937) grenade. Weight : 1 lb. Delay : 4–5 seconds. Thrown.
Right : Model 99 (1939) " Kiska " grenade. Weight : 10 oz. Delay 4–5
seconds. Thrown, or fired from a rifle cup discharger.

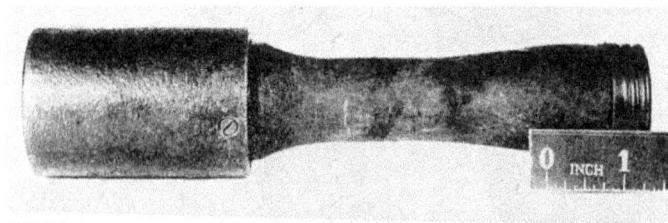

PLATE 7.—HE Stick Grenade.
Weight : 1 lb. 3½ oz. Overall length : 7·87 in. Delay : 4–5 seconds.

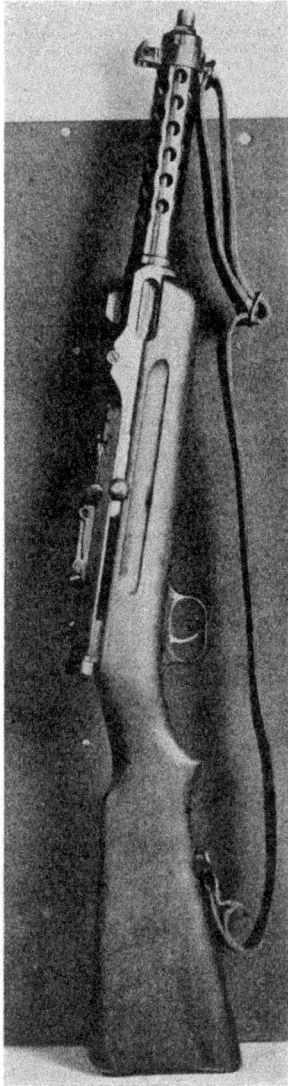

PLATE 8.—7·63 mm (·30 in) machine carbine (Solothurn).
Weight (without magazine) : 9¼ lb. Length : 32¼ in. Feed : 32 round box magazine. Cyclic rate of fire : 700 rpm.

PLATE 9.—6·5 mm (·256 in) LMG TAISHO " 11 " (1922).

Weight : 22½ lb. Overall length : 43½ in. Feed : Hopper, loaded with six 5 round clips. Cyclic rate of fire : 500 rpm. Ammunition : semi-rimless. Sighted to 1,500 metres.

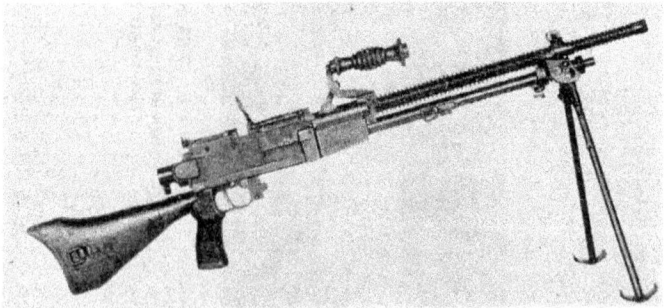

PLATE 10.—6·5 mm (·256 in) LMG Model 96 (1936).

Weight : 20 lb. Overall length : 42 in. Feed : 30 round box magazine. Cyclic rate of fire : 550 rpm. Ammunition : semi-rimless. Sighted to 1,500 metres.

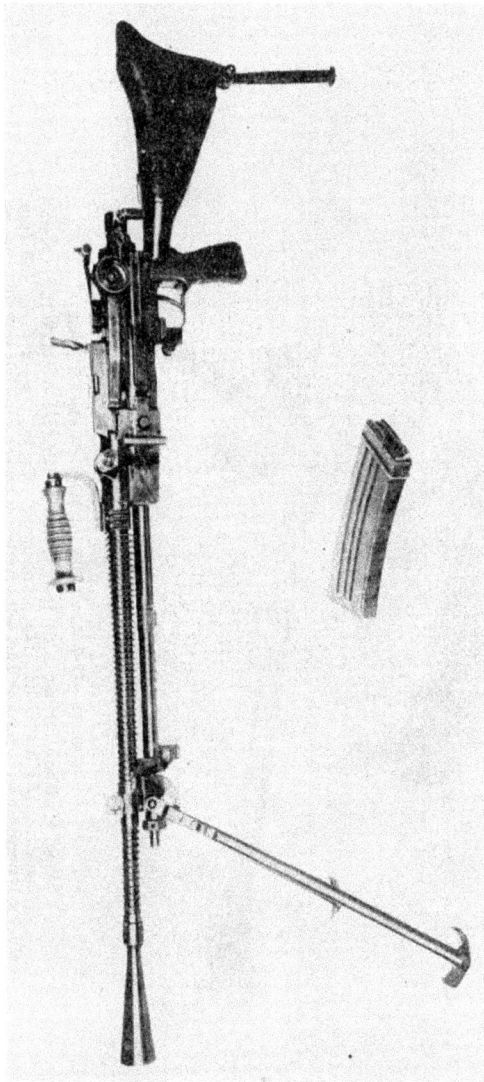

PLATE 11.—7·7 mm (·303 in) LMG Model 99 (1939).

Weight : 22 lb. Overall length : 42 in. Feed : 30 round box magazine. Cyclic rate of fire : 800 rpm. Ammunition : rimless. Sighted to 1,500 metres.

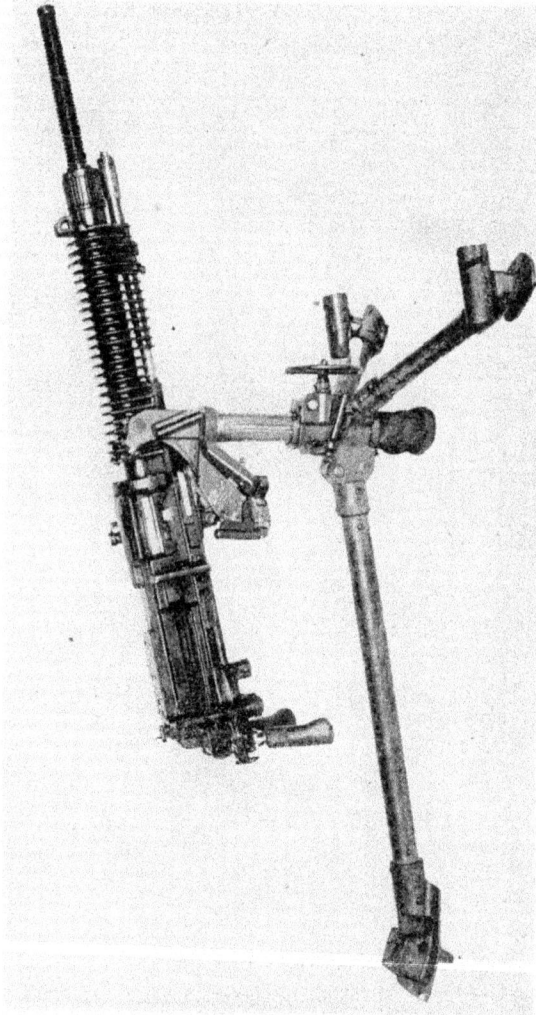

PLATE 12.—7·7 mm (·303 in) MMG Model 92 (1932).

Weight without tripod : 61 lb. Weight with tripod : 122 lb. Overall length : 45½ in. Feed : 30 round strip. Cyclic rate of fire : 450 rpm. Ammunition : rimless and semi-rimless. Direct sighting to 2,700 metres. Dial sight provided for indirect fire.

PLATE 13.—20 mm (·79 in) A tk rifle Model 97 (1937). .

Weight in action : 140 lb. Overall length : 82¼ in. Feed : 7 round box magazine. Sighted up to 1,000 metres. Provided with a shield and carrying handles.

SECTION 9
MILITARY BRIDGES

SECTION 10

DEMOLITION EQUIPMENT

The Japanese Army makes use in general of small types of demolition charges and equipment, which can be carried and used by the individual in difficult terrain.

There is a distinct trend in design to imitation of German models, except that the Japanese charges do not contain a standard threaded socket.

The explosives generally used are picric acid, TNT, and plastic explosives; picric acid is the most common and great care is taken to protect this from moisture, by use of waxed paper and zinc casings.

Igniters encountered include friction, pressure (concussion), pull (with and without delay).

Plate 1.—100 gm (0·22 lb) Demolition Charge

This charge, measuring 2 × 2 × 1 in is wrapped in waxed paper. It may or may not be provided with a hole for a detonator. The explosive is cast picric acid.

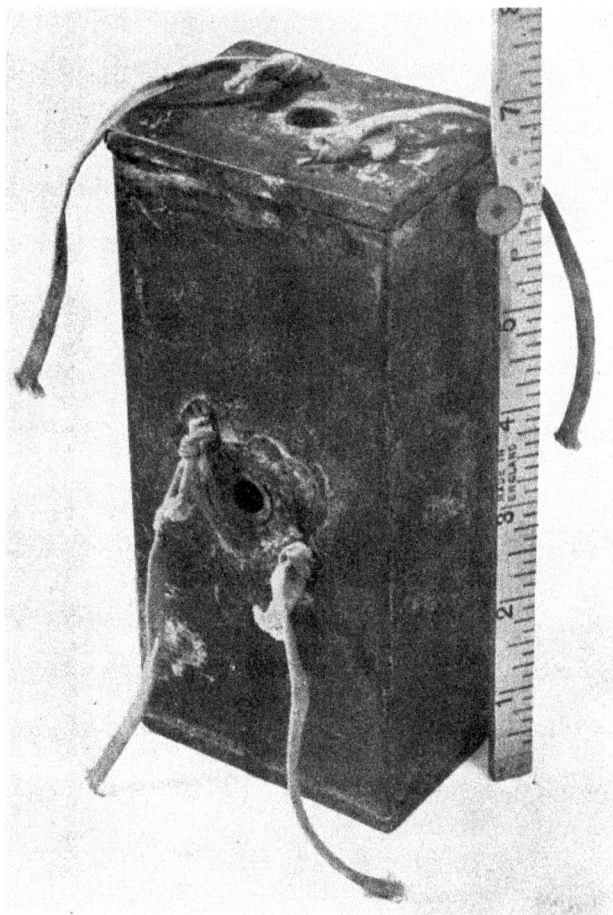

Plate 2.—1 kg (2·2 lb) Demolition Charge

This charge is contained in a zinc case (25 SWG) measuring $2\frac{7}{8} \times 2\frac{1}{8} \times 8$ in, and having two brass tubes for detonators.
Tapes or silk thread may be used, presumably for retention of the detonator.
The total weight of the charge is 1,300 gm (2 lb 14 oz).

170

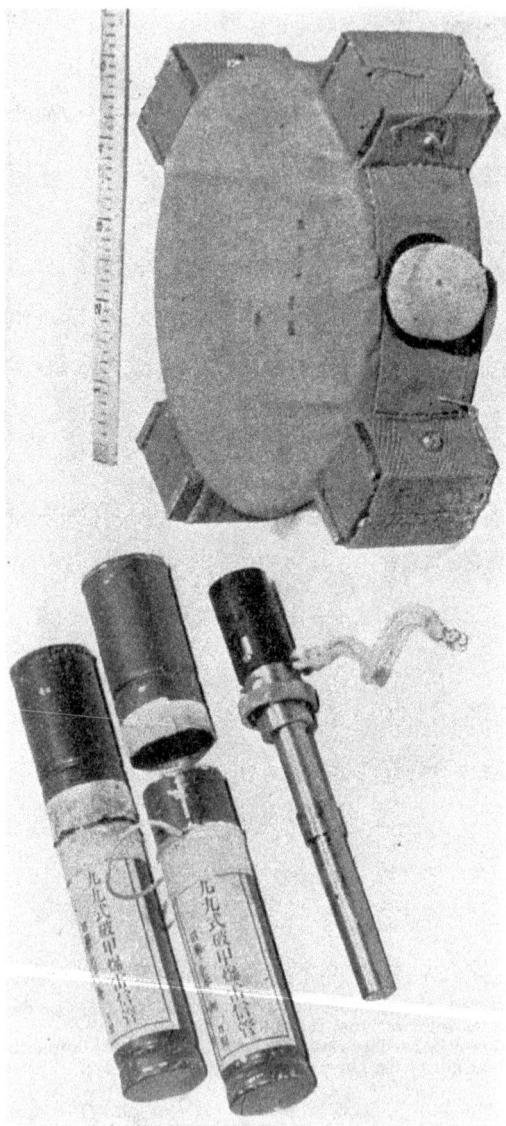

PLATE 3.—Magnetic Demolition charge, Model 99 (1939). (Hakobakurai).

This charge weighs 2 lb 11 oz complete and contains 1 lb 1 oz of cast TNT in eight specially shaped blocks. A complete igniter set is carried in a separate metal case. When the safety pin is withdrawn the end cap of the igniter is given a sharp knock to ignite a percussion cap and powder delay train. The delay period is 5—6 seconds. The charge is reported to penetrate 20 mm armour plate.

The charge has to be placed on the armour; throwing is not a successful operation owing to weakness of the magnets.

SECTION 11

MINES

No standard anti-personnel mines have been used by the Japanese Army so far. Booby traps and improvised mines making use of the various types of hand grenade in conjunction with trip-wires, pull and pressure or concussion methods of ignition have, however, been profusely employed on all fronts.

The standard anti-tank mine used was the Model 93 "tape measure" mine (*see* Plate 1), and although reports have been received of other types they have so far not been substantiated. Improvised mines have also been employed against tanks with varying success.

An account of the Magnetic Demolition Charge, Model 99—sometimes referred to as an A tk mine—will be found in the Section on Demolition Equipment.

The anti-boat mine, Model 98 (hemispherical) has been extensively used by the Japanese as an obstacle to beach landings.

PLATE 1.—A per/A tk Mine, Model 93 (1933).

The mine is 6¾ in in diameter and weighs 3 lb complete. The weight of explosive is about 2 lb. When the igniter is fitted with an A tk shear wire the firing pressure is believed to be about 250 lb; the shear wire for A per use is reported from Australia to fail at 70 lb.

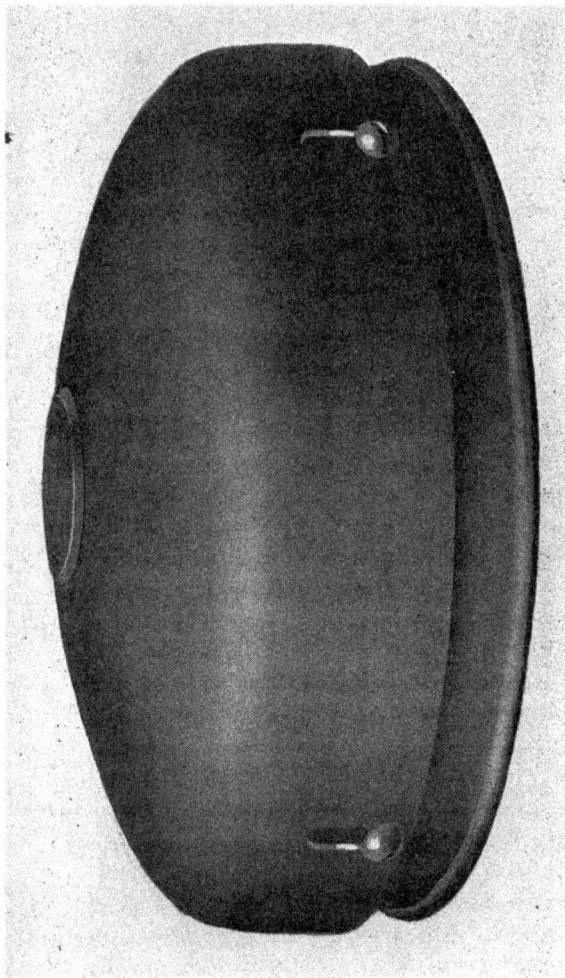

PLATE 2.—A per/A tk Mine, Dutch model (PW2—41).

This mine, originally Dutch equipment, was found in Guadalcanal. It measures 8¼ in diameter by 3¼ in high, contains 5¼ lb of cast TNT and weighs 9¼ lb. The firing load is believed to be about 50 lb.

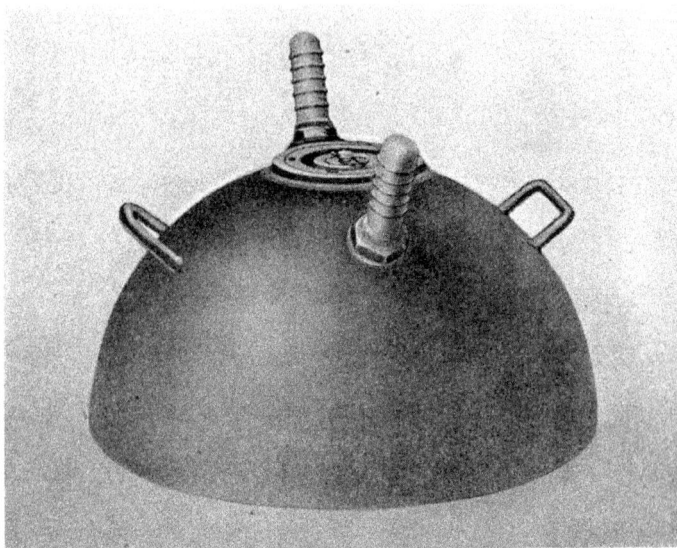

PLATE 3.—Anti-boat Mine, Model 98 (1938).

This mine is 1 ft 8 in in diameter, 10⅝ in high and weighs 106¼ lb. with
an explosive filling of 46¼ lb. The mine is provided with an arming
device in the top which consists of plunger contact assembly, electric
detonator and primer. It is initiated through the crushing of either of
the two lead-alloy horns containing a phial of electrolytic solution which
passes into a cell to complete the electric circuit. In addition to being
laid against landing craft these mines have been found above high water
mark with wires tied to their horns for initiation.

SECTION 12

FLAMETHROWERS

While there have been unconfirmed reports of a Japanese heavy tank mounting a flamethrower, the only flame warfare weapons so far encountered have been the light portable models which are carried on the operator's back.

The following types are known to exist :—

Model 93.
Model 93 modified.
Model 100.

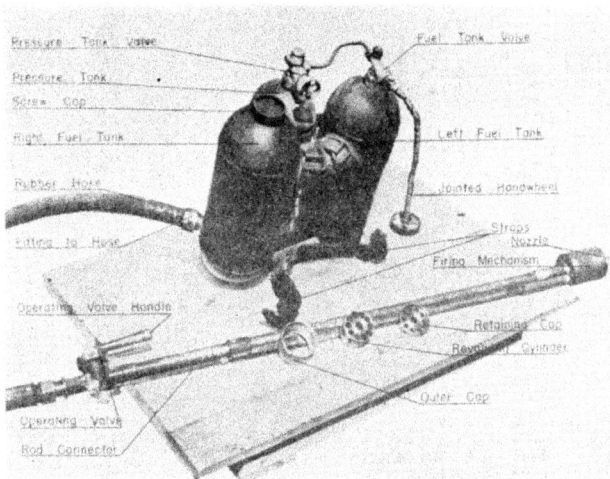

Plate 1.—Light Portable Flamethrower, Model 93 (1933).

This equipment, with a charged weight of 55 lb and a capacity of $3\frac{1}{4}$ gall, fires an oil type fuel to a maximum range of 30 yd. The duration of continuous discharge is 10—12 sec.

The fuel is ignited by flash from a blank cartridge, ten of which are loaded in the revolving cylindrical chamber at the forward end of the flame gun. The firing mechanism is actuated by operation of the handle which also controls fuel emission.

The model 93 modified flamethrower has a shorter flame gun and a modified cartridge chamber mechanism.

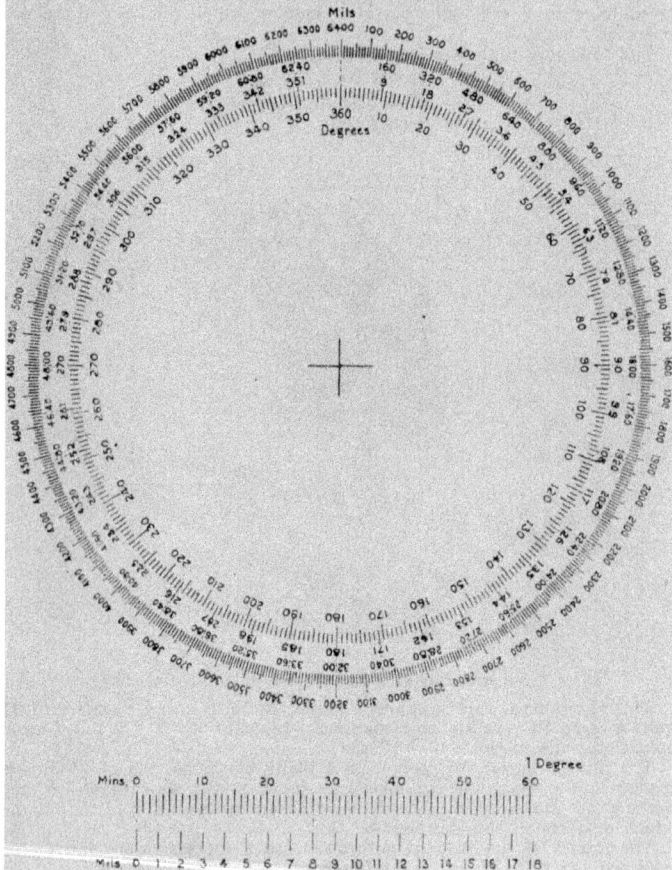

CONVERSION DIAGRAM
(Cont.ᵈ Mils to Degrees & Minutes)

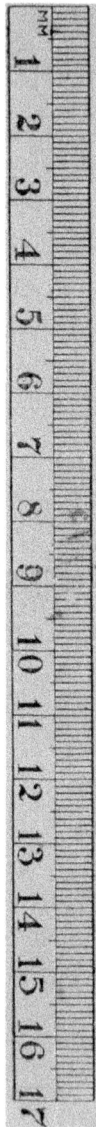

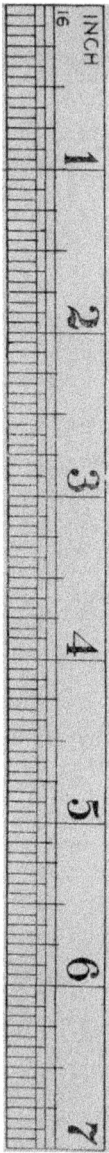

MM 1 2 3 4 5 6 7 8 9 10 11 12 13 14 15 16 17

INCH 16 1 2 3 4 5 6 7

TRAINING MANUALS, TEXT BOOKS AND INSTRUCTIONS

The backbone of all successful armies is its training and tactics. The Naval and Military Press publishes many such manuals of instruction – all perviously long out of print . So, whether your interest lies in the infantry and cavalry tactics of the earliest regiments of the British army in the 18th century, or the weapons manuals and firing instructions of 20th century warfare, the Naval and Military Press has the right book for you.

www.naval-military-press.com

MINES AND BOOBY TRAPS 1943

This is a War Office pamphlet, issued mid-war, in 1943. Its purpose is to introduce sappers to mines commonly used by the British Army – and how to deal with similar devices set by the Germans. The devices described and illustrated cover British anti-tank; grenade; shrapnel and assorted booby trap switches. Enemy mines are covered in chapter 2 with anti-tank, Teller mine types; French anti-tank; Hungarian; anti-personnel German and Italian; and igniters.This is a concise but comprehensive guide for British Army sappers in the art of demining or mine clearance.

9781474539395

THE .303 LEWIS GUN

Illustrated with good clear line drawings this 1941 weapon guide tells the Home Guard Volunteer how to use the 303 Lewis Gun effectively against the invading enemy.A reprint of an original handbook for the .303 Lewis Gun, that was first published in 1941. This book is a practical guide to the handling and maintenance of this iconic weapon.In the crisis following the Fall of France, where a large part of the British Army's equipment had been lost up to and at Dunkirk, stocks of Lewis guns in both .303 and .30-06 were hurriedly pressed back into service, primarily for Home Guard use. Full of fascinating information, this book taught the user the guns capabilities and all he needed to know about maintenance and combat use. Number 2 in the wartime Nicholson & Watson "Know Your Weapons" series, that offer all the important information in a more vivid style than an official publication. Illustrated with good clear line drawings.

9781474539456

ANTI-TANK WEAPONS
Smash The Tank

An insight into the amateur side of World War 2. Diagrams illustrate the main points and the devices, such as the Thermos Bomb;Phosrhorus Bomb;Sticky Bombs; that could be cobbled together from household items are described.This pamphlet was available to the Home Guard and describes the German tank and how to destroy it. It is an early War publication c1940, dealing with the light tanks used by the Germans, also the author gives examples of anti-tank actions in the Spanish Civil War, in which he took part. I'ts is a fascinating look at the "enthusiastic" approach to killing tanks.

9781474539449

TANK HUNTING AND DESTRUCTION 1940

The stated object for the distributing of this War Office manual was as "A guide and help to troops who have the determination and nerve to destroy tanks at close quarters". Intended for fighting on home soil after the very real possibility of a full German invasion, "Operation Sea Lion", this is a remarkable if somewhat naive snap shot of Britain state of preparedness,in her most dangerous hour.

The contents details Tank hunting, Tank characteristics,Tactical action,Road blocks,ambushes Ect,also includes an interesting appendix on Molotov Cocktails, and materials on other ways to destroy tanks.

9781474539401

TROOP TRAINING FOR LIGHT TANK TROOPS NOVEMBER 1939

Very early War tactics pertaining to various aspects of training with and employing armour in the British Army. Covering in concise detail that which a Light tank crew needed to know to be effective in action. In the early years of the war, Germany held the initiative. German forces used Blitzkrieg tactics in France in 1940, making full use of the speed and armour of tanks to break through enemy defences. It was clear that German tank tactics had evolved during the inter-war period. By contrast, Britain and the Allies were playing catch-up.

9781474539302

JAPANESE WEAPONS ILLUSTRATED
September 1944

This period 'Restricted' laced binding manual was intended to be an aid to the identification of Japanese Army equipment, with sections covering: Tanks, both two-man, Tankette, light and medium; Armoured Cars; Self-Propelled Guns; Anti-Tank Guns; Artillery; Anti-Aircraft Guns; Mortars & Grenade Dischargers; Small Arms; Flamethrowers etc. Produced one year before the surrender of Japan, this work gives a good overview of the weapons the allies would find, fighting an army that despite being on the back foot, was still capable of stiff resistance in an almost entirely defensive role..
9781474539432

NOTES ON THE GERMAN ARMY-WAR
December 1940

An early war 393-page 'Notes' periodical manual from December 1940. It is a detailed review, for use in the field. The manual looks at every aspect of the "Blitzkrieg" German Army (and, to some extent, the Air Force) and gives details as known at the time.

It covers the fighting arms and the services behind them – tactics, organisation, weapons and equipment. It usefully also includes a colour section on uniforms and insignia, a black-and-white plate section of small arms, infantry support and anti-tank weapons, artillery and AFVs. A series of pull-outs related to the text covering tanks etc. are also reproduced.

This is an important first-class picture of the complex fighting machine that was the German Army at the end of the campaigns of 1940, only six months before the invasion of Russia.
9781474539203

GERMAN MINES AND TRAPS

Mid-1940 War Office manual with details of German mines, both the Teller and S-mine (Bouncing Betty) are covered, with techniques for disarming. Good clear full-page line drawings give both practical and technical information. Highly recommended because of the illustrations, which show how these devices worked and the components.
9781474535809

NOTES ON ENEMY ARMY IDENTIFICATIONS ITALY
October 1941

This period handbook was published to give British military personnel a better understanding of the principal characteristics of both the Italian army and the Black Shirt Militia under active service conditions , it is dated October 1941.

It begins with a description of distinctive branches, or specialities, the most characteristic of which was the arm of the Royal Carabinieri, a semi-military body occupying, historically, the senior position in the Army. Other specialities included the Grenadiers of Sardinia, the Bersaglieri, the Alpini and the San Marco Marine Regiment

The handbook then goes on to show, in order, the organisation of Command and Staff, of formations (corps and divisions) and of the arms and services; services, supply and transportation; ranks, plates (many in colour) cover uniforms, insignia, medals and decorations; armament and equipment and a chapter on the Air Force, There are chapters on tactical doctrine and principles of employment, on permanent fortifications, camouflage and abbreviations. Finally there is a brief index.

9781474539746

MANUAL OF GUERILLA TACTICS
Specially Prepared And Based On Lessons From
The Spanish And Russian Campaigns

One of the excellent, concise Bernards Pocket Books, intended to show members of the Home Guard and the regular forces that war is not conducted in a gentlemanly way – it is kill or be killed.

9781474539463

THE OFFENSIVE OF SMALL UNITS
September 1916

This is a periodical tactical manual from 1916, it focuses on the manner in which the French organised and executed their attacks and counterattacks . Summarised from the French, it lays out the process by which to operate in attacks on the German trenches. Focused purely on the operation of infantry, the purpose of this British translation is to give small infantry units the benefit of the French experience in regard to the best methods of combat, in offensive operations.

9781474537971

TRENCH WARFARE
Notes on attack and defence, February 1915

This important period manual was published in early 1915 when hope of a quick ending to the war disappeared, and trench warfare had begun to dominate the Western Front.

The manual strives to instil an offensive spirit and gives practical examples on: Close quarter, local, methods of successful warfare, and German attacks. The salient points to gather were preparation and co-operation between artillery and infantry, and that the capture of trenches is easier than their retention. Two plates illustrating tactics complete this official publication.

9781474539807

Ministry Of Home Security
OBJECTS DROPPED FROM THE AIR 1941

An illustrated Official and confidential publication, covering the many and varied types of objects that were falling from principally German aircraft during the Second phase of the blitz, including high explosives,incendiary bombs and small arms ammunition. Complete with 8 page addendum.

9781783319541

THE MUSKETRY INSTRUCTIONS
FOR THE GERMAN INFANTRY 1887
(Schiessvorshrift fur die Infanterie)
Translated for the intelligence Division War Office

Translated for the War Office by Colonel C W Bowdler Bell

A facsimile that includes the supplement for the German Infantry for 1887. Musketry exercises were intended to give the infantry instruction in shooting, to make effective use of their firearm in battle. As such the manual shows important details designed to make the infantry soldier battle-ready by the end of his first year of service. Instruction is subdivided into Preparatory exercises; Target practice; Field firing; Instructional firing; Inspection in musketry; Proving the rifle M/61.84 and revolver M/83. Many black powder weapons were still used, mainly for training purposes, up to end of the First World War.

9781783313631